WHAT NEXT ? !

BOOKS BY THE SAME AUTHOR

Science, Ethics and Politics

The Crazy Ape

WHAT NEXT?!

With a Preamble by

Daniel Webster

Albert Szent-Györgyi

M.D., Ph.D., Nobel Laureate

PHILOSOPHICAL LIBRARY
New York

PHILOSOPHICAL LIBRARY, Inc.

15 East 40th Street, New York 16, N.Y.

Library of Congress Catalog Card Number: 73-136016

SBN 8022-2045-2

Manufactured in the United States of America

CONTENTS

WHAT NEXT ? !

I. PREAMBLE

"Where is it written in the Constitution, in what article or section is it contained, that you may take children from their parents and parents from their children, and compel them to fight the battle of any war in which the folly and wickedness of the government may engage itself?

"Under what concealment has this power lain hidden, which now for the first time comes forth, with a tremendous and baleful aspect, to trample down and destroy the dearest right of personal liberty? Who will show me any Constitutional injunction which makes it the duty of the American people to surrender everything valuable in life, and even life itself, whenever the purposes of an ambitious and mischievous government may require it? . . .

"A free government with an uncontrolled power of military conscription is the most ridiculous and abominable contradiction and nonsense that ever entered into the head of man."*

---

*Daniel Webster in House of Representatives, January 14, 1814.

Among brigands the silent
is an accomplice.
(M. Babits)

II. INTRODUCTION

After my last book, *The Crazy Ape,* appeared, I found that in discussing its contents with my young friends a common complaint was voiced: "It does not tell us what to do, or how." I have written the present book to correct this shortcoming. To make myself clear I will have to enlarge upon a few points discussed in my first book, which makes a moderate amount of repetition unavoidable.

III. THE THREE PERIODS OF HISTORY AND THE ULTIMATE BIOLOGICAL ABSURDITY

If we want to stay alive and do not want to join the Dinosaur, we have to reshape all our institutions, as well as our political, social and ecological thinking. To understand this we must look at our problems from a broad historical perspective.

Man being small and nature being big, man's life depends on his relation to nature, on the measure at which he understands nature and can use her forces to his advantage. From this point of view human history can be divided into three major periods. In the first man used his primitive daily experience to shape primitive tools or weapons. With Lewis Mumford we can call this period the age of "paleotechnics."

In the sixteenth century, during the time of the Renaissance, man found a new way to approach nature by measuring, calculating, asking questions from her — that is, doing experiments. Galileo, one of the first to use this approach, was soon followed by a rapidly increasing number of others, now called "scientists." This new approach led in the nineteenth century to the construction of more complicated instruments and machines which could ease man's toil. Machines led to the Industrial Revolution, which greatly improved the quality of man's life. We can call this period (again, with Lewis Mumford) the age of *"neotechnics."* The science of this period reached its peak with Newton, Darwin, Mendel and Pasteur. It brought no new factors into our world but clarified many of its internal problems. Apples fell even before Newton; what we did not know was that our globe circles around the sun for the same reason that apples fall. We knew before Darwin that there were many species; what we did not know was that these were all interrelated and developed from one another.

The third and last major period started twenty-five years ago. The earlier periods needed centuries in which to gain a shape. This last one started on August 6, 1945 at 8:15 A.M. It started with a bang, the holocaustic bang of the bomb at Hiroshima, the first atomic bomb used to destroy man's life and creation. In the matter of a few seconds it leveled a big city, one that was built over many centuries, and wiped out about 100,000 human lives. This event announced to man that life would never be as it was before. This last period had its roots in discoveries made at the turn of the century, which showed that the ultimate units of our world were not rocks, caves, trees, bears and the like, as man has thought before, but quanta, atoms, atomic nuclei, radioactivity, electrons and electromagnetic radiation. These are the units out of which the cosmos is built. This new knowledge made man the master of cosmic forces, and at the same time made our terrestrial rules obsolete. No longer was man able to live simply according to earthly rules he had developed through centuries of trial and error.

He was now made conscious of, and required to live with, the cosmic rules and forces.

It is interesting to reflect, for a moment, why this new period broke in on us so suddenly. Basic discoveries always have a certain latency period before they begin to transform our life. The discoveries made at the turn of the century needed about three decades to leave their mark on human life, but just when they matured to fruition mankind suffered an intellectual blackout, being involved in a gigantic struggle, the war against fascism. So we did not notice the change. All the same, cosmotechnics already had a decisive role in this war with one of its instruments, the radar. Had the war lasted a year longer, cosmotechnics would have taken over altogether. Had Hitler developed the atomic bomb first, nothing could have stopped his world conquest. Our bomb was a late-comer in the war and was dropped after Japan had already caved in. All the same it helped to put an end to the struggle and spared a great number of American lives.

The age of cosmotechnics dawned so suddenly that it tore mankind in two: into those born and brought up in the pre-cosmic period, and those born and raised after. It tore families in two, creating the "generation gap," parents belonging to the pre-cosmic, their children to the cosmic age. The political and social turmoils of our day are due to a great extent to the fact that nearly all our leaders come from the earlier pre-cosmic period. Their minds have frozen up with the rules and feelings of a past age: Their ways may have worked well in their own period, but are disastrous in ours. They have ruined, to a great extent, the richest, most vigorous and progressive country in the world — rich not only in material goods but equally rich in intellectual treasures. The leaders who have built their mental framework and established their rules of conduct and thinking in the earlier period may sincerely be striving to do their best, but they are doing their worst, starting wars and extending them, neglecting all our real priorities, resigning all our claims for leadership, degrading the U.S.A. by making it the world's policeman instead of its moral leader. They are leading all mankind towards extinction.

In *The Crazy Ape* I quoted a few numbers from J. Platt to illustrate the change from "neotechnics" to "cosmotechnics," and I would like to reflect here for an instant on the meaning of these numbers; but before doing so I must explain two words often used in science: "qualitative" and "quantitative." When I say that two things are different quantitatively, I mean that they are different in size only, while by different qualitatively I mean that the two things are different in kind. A rat and a rabbit are different not only in size, but also in kind: they are different animals, different qualitatively. If I would make a rat as big as a rabbit, it would still be a rat. However, in extreme cases, quantitative differences become qualitative. If I were to make my rat a million times bigger, make it weigh 500 tons instead of 500 grams, then it would be a terrific new monster, something entirely new — different qualitatively.

One of the most important factors in modern life is communications. The speed of communications has increased in today's world ten-millionfold. I can watch, sitting at home in slippers, not only how a man walks on the moon, I can even hear him talk. What does this mean? It means that the world of today is qualitatively different from the world of just a few years ago. The globe has, by virtue of communications, become fused into one single unit; the result is that we now live in a world in which a virus born in Hong Kong can within days ravage New York, in which the killing in Vietnam makes my street unsafe, and the bombing in Vietnam makes bombs go off at home. In earlier periods our planet was cut up into small, isolated, national units which could quarrel amongst themselves without any dire consequences for the rest of the world. This, of course, is no longer the case; in today's world any local quarrel can eventually lead to the extinction of us all.

The most modest number I noted in *The Crazy Ape* was the speed of travel, which increased only a hundredfold. What does this number mean? Distance, in itself, has no meaning for us; the question is how fast we can cover it. A mile for a snail is a longer distance than a trip to the moon for an astronaut. So if the speed of travel increased a hundredfold, this means that distance has shrunk to 1/100 of what it was before. We did see this happen before our eyes. Lest dealing in such figures seem too abstract, here is an example that many of us have witnessed: the Atlantic, which thirty years ago protected our country with its endless expanse, has today shrunk to a ditch across which we can travel in a few hours, and over which, in a matter of minutes, aimed missiles can soar. Similarly, our globe has shrunk to a little ball.

What makes this situation so dangerous is that while the

world was contracting, the destructive power of our weapons increased a millionfold, and we find ourselves standing on our shrunken little spaceship with weapons in our hands a millionfold more powerful than ever before, powerful enough to exterminate us all at a moment's notice. Here we stand, with these weapons in our hands and with our petty national hatreds, jealousies, distrusts and lust for domination in our hearts. If we want to stay alive we have to throw away not only these weapons, but also our petty nationalism and all the things this represents, replacing them with a wider human solidarity, replacing force with decency and equity. Our leaders, inherited from the old world, from the old way of thinking and the old way of problem-solving, cannot see this. They still think they are twice as safe and strong with twice as many bombs, whereas the fact is that they are merely twice as close to the brink. There are only a few in the older generation of leaders who have understood the meaning of the new, cosmotechnic age.

All this has led to the ultimate absurdity which makes one wonder about the mental state of our leaders. Two radar systems are looking at one another, that of the Soviet and that of the U.S.A. If the one, say the U.S.A.'s system, sees missiles coming, it has to give the order "fire" to have our missiles in the air before the Soviet missiles arrive and interfere with our firing. As explained by Herbert York in his book *Race to Oblivion,** there will be 15 minutes left to evaluate

*Simon and Schuster, New York, 1970.

the situation and make the final decision. There is no man in the world who can evaluate such a complex situation in such a short time, and there will be but one thing left: to fire, and exterminate mankind. Machines are machines, but in one way they are human: they can make mistakes. Nature needed three billion years to build man and we spend all our money to be able to wipe him out. This is the greatest biological absurdity, as pointed out by George Wald.

We have spent, since World War II, one trillion dollars, $1,000,000,000,000, on safety, to buy a free ticket to extinction, to have our fate, and that of our children and the whole of mankind, entrusted to two machines run by a few people whom we have no special reason to trust. With this sum of money, we could have raised all mankind to a higher level. This sum is too big to be imagined, and we spent it while half of the world's children went hungry to bed without sufficient protein to build healthy minds and bodies. A crime, indeed. We, the U.S.A., have spent about twice as much on "safety" as the Soviets so it is fair that we should have some extras. We have them. We cannot go out on our streets at night without the fear of being robbed, and we are made to feel unsafe even at home. And we are still continuing on the same senseless course, building ABM, multiplying our atomic warheads by MIRV, and our President still vetoes millions for health and education as inflationary, while spending dollars by the billion on increasing and improving our arsenal, though we already could overkill every human being ten times over, and it is difficult to kill anyone more than once. We also have to sink our excess poisons into the sea, destroying the sea life which feeds us. This is what we call a "Country in God."

Who is responsible for this ultimate absurdity? Lunatics, criminals, or simply leaders belonging to and earlier age who look into the past for the future?

The perennial and overriding question that derives from all this is: what can we do to alter our suicidal course? In a very strong sense this is the question that has divided our country during the past decade. As more and more people, especially young people, came to the realization that we could not continue to apply the solutions of the precosmotechnic world to the problems of the new world and expect to survive, they began to seek the proper alternatives. They met with the inevitable resistance of those who cannot think in new terms or are bound by their vested interests. The solution is simple: exchange the leaders of the past age with new ones who have their roots in the new age, and retain from the older generation only the exceptional ones who were able to adapt.

But how can such a change be made? Do we have to destroy the old world before we can build a new one? Certainly not. We Americans have a wonderful document in our Constitution, which allows us to make any change smoothly, without violence or bloodshed. Indeed, the Constitution provides for change and makes it possible to effect change simply by dropping a piece of paper into a box — our vote into the ballot box. Of course, this process is not as simple as it sounds because the majority of voters still belong to the past age and will probably continue to cast their votes for the old order. So we must use all our intelligence to achieve a majority for a better world. Modern science has enabled us to create a new world, without pain, hunger, toil and disease, with heretofore undreamt of wealth, dignity and happiness; but it has also enabled us to wipe ourselves off the surface of our lovely little planet. The choice is ours.

IV. THE NATURE OF MAN AND THE NATURE OF WAR

We are told that man is a ferocious, bloodthirsty animal, and wars are the natural consequences of his nature. We are told this over and over again. This is a crucial point. If it is true then we must resign ourselves to war and to extinction. According to historians, during the five thousand years of our history there were only three hundred years without war. Some of our wars lasted thirty years and led to the depopulation of whole continents. Does this not mean that we are bloodthirsty killers?

If we are not bloodthirsty killers, all efforts are being made to make us into ones, to enable the establishment to use us, make us fight and die for its schemes or interests.

Educational systems are always instruments of the establishment, and try to make us into its instruments. One of the methods to achieve this end is endless cramming, instead of inculcating a critical mind. I remember the pictures on the wall of my class at primary school, which soaked unconsciously into the young minds. They were all pictures of battles in which always we galloped to victory over the dead bodies of the enemy. The false impression that only the enemy is killed in a war is corroborated by life, because it is only the survivors who return with their romantic stories. If only once those could come home who died ignominously in dirt, history would take a different course. The history I learned in school was the glorification of mass murder, glory being proportional to the number of people killed or cities destroyed. For twelve years, I was taught that the peak of glory was to die for the King. Later I learned that the "king" was but a gang of militarists, led by Count Berchthold, Secretary of the Army, who forced our old emperor, Franz Joseph, to sign the declaration of war with falsified documents.

Having served five years myself in World War I, I have a clear recollection of it. There was no enthusiasm, no joy of going out for a kill, in spite of all the drinks, bands and flag waving. We were sad and frightened, and if I look around myself today, I see no bloodthirsty animals, I see only human beings. Man is an anthropoid ape and the ape is a shy animal. We are shy apes and not bloodthirsty killers. All animals, the tamest ones, like the deer, attack when cornered. The trick the army uses to make us into killers is to put us first through a drill, to wipe out one personality, then jockey us into a position where we have a choice only between killing and being killed, shooting or being shot. It is a successful method, sometime too successful.

Next to the parental house, and the school, the Government is the most powerful educator, shaping the people to its needs.

It has a free access to mass media. The Government is, as told by Chet Huntley, the former NBC news broadcaster, a jumble of agencies pursuing their own interests. The military-industrial complex is by far the most powerful of them and is more or less fused with the Administration. They need killers for Vietnam, and are thus depreciating cultural and human values, making us forget our real priorities, brutalizing and dehumanizing, to ride to the next election on the wave it wiped up against youth and intellectuals.

My conclusion is that wars are not a necessary consequence of man's nature but the result of the machinations of individuals and their gangs seeking their own glory or interest. This leads to the important conclusion that war can be prevented simply by denying all support to warmaking machineries and hawkish political groups or people.

The question arises that if man really wants peace why does he give his allegiance so often to those who make war, spread hatred and hostility? Whence this support for hawks? Does it not prove that at the bottom of our hearts we do want war? Such appearances can be very misleading and the probable answer to this question was given two thousand years ago by *Juvenalis,* the Roman satirist (60-140 A.D.), according to whom herds of sheep always beget governments of wolves. One might translate this into contemporary language as: flocks of doves beget governments of hawks. As there are always a few rotten apples in a barrel, so there are always hawks among the doves. They are out for power and fight their way to the top. Once on top of their nation, they want to get on top of the world and make war. If doves happen to elect a dove as a leader, the hawks assassinate him; whereas doves do not assassinate hawks, so they end up with hawks as leaders, who then try to turn the rest into hawks.

Naturally, this may be not all there is to it. There seems to be a change in the mind of man when he gets power, for "power corrupts." As I pointed out in my earlier book, one death is a tragedy, thousands of deaths are statistics. Perhaps human nature is such that if a man looks down on his fellow men from high up, he begins to think in terms of statistics and readily sacrifices the lives of his fellow men for his schemes and then justifies the slaughter of his own people by the number of other human beings killed, whom he calls "enemies."

In our weekly report on the number of slaughtered G.I.'s in Vietnam, the account is put straight by the greater number of S. Vietnamese and ten times greater number of N. Vietnamese killed. The "body count" is kept as a football score, showing that we have won by causing more deaths than we have suffered.

War, in the age of cosmotechnics, has its specific features. Earlier, if two nations quarreled they could fight it out between themselves. Whatever the result, the war was no danger for the race, for one of the two parties always survived. War in the cosmic age is different: the bystanders will be killed too, and all mankind will be wiped out. No powers have the right to do this just to settle their private quarrels. The planet is not their private property. It is the common property of all mankind, and so are its atmosphere and its oceans, and nobody has the right to pollute them. The two contending powers both confess to democratic or socialistic principles according to which big and small nations have equal rights to exist; but at the same time they stockpile thousands of bombs with which to wipe us all out with the final and fatal poisoning of the atmosphere to settle their differences, and poison the oceans with the surplus nerve gas which they produced in their insanity. Why do the smaller nations not get together and tell the great powers to destroy their bombs and thereby write a *habeas corpus* for mankind? Together, the smaller nations are still the majority. What an incredibly callous arrogance it is on the part of the super powers to threaten their existence! Do the governments of big powers know no decency, to think they have the right to do what they please because they are stronger?

I do not imply that all this is a result of ill will on the part of the large nations. There may be a great deal of governmental ignorance and callousness, but what is at the bottom of it all is mainly that the cosmotechnic age has overtaken us. It has introduced new dimensions, and in order to cope with them we have to discard the old world and build a new one.

As the atmosphere is the common property of man, so is the ocean. While our navy pretends to be the guardian of the freedom of the oceans, this very minute, while I am writing these lines, we prepare to scuttle a boat in the ocean, a boat filled with lethal nerve gas, packed into thin metal containers which soon will deteriorate, poisoning the ocean which feeds us, killing all life in it. Earlier we dumped 17,000 similar bombs into the ocean. While we pretend to fight pollution, we go on fabricating the most deadly weapons to destroy life, to poison water and air and kill millions of acres of virgin forest with our defoliants.

It is time to place a sign on our little space-ship:

PLAYING WITH ATOMIC BOMBS
AND
POLLUTING
IS STRICTLY FORBIDDEN.
MAXIMUM OCCUPANCY 3½ BILLION
PERSONS

V. THREE WAYS TO SUICIDE

According to the majority of leading scientific thinkers, three processes are rampant in the world today, each of which is certain to exterminate us if allowed to go unchecked: pollution, overpopulation, and armaments, a fourth may lie in the senseless squandering of the very limited resources of the globe. Pollution and overpopulation work slowly and are natural consequences of the past. We have recognized these dangers and, hopefully, we will be able to do something about them.

The situation is different with armaments — nuclear and conventional. They are still proliferating and can wipe us out any moment. There is no real need for them. Hundreds of atomic missiles are ready to be fired against us, while we have twice as many, enough to kill the entire Soviet people three times over. As we build MIRV we multiply their number; our submarines are in dry dock where their atomic warheads are being changed into these "Multiple Independently Targetable Re-Entry Vehicles," which will multiply our destructive capacity tenfold, enabling us to over-kill thirty times instead of three. Our capability to produce fallout, combined with that of the Soviet, is sufficient to wipe out all mankind. We have the equivalent of 15 tons of high explosive (TNT) for every human being. In addition to this we have all sorts of poisons, each in sufficient quantities to kill every living creature.

The way the Army works is this: firstly, it hires half to two thirds of all scientists, taking them away from work on human welfare, making them search for new weapons. Once they find a new method for mass slaughter this is subjected to the "worst case analysis," supposing that our enemies also know about it and are in full scale production. So, to be safe, we also have to go into full scale production. Once we produce these new weapons our opponents also have to produce them. Once they produce them we must produce them still faster, to remain ahead, and so forth. With every new weapon or improvement we are closer to the brink instead of being safer. The trouble is that one cannot have one's cake and eat it too. If we spend our money on armaments, on the instruments and organization of killing, we must neglect all our real priorities, human interests. This leads to a gradual brutalization and dehumanization, which speeds up the arms race. At present we are directing 48.4 cents of every tax dollar to military expenditure. Another 16.4 cents goes for the expense of past wars, so altogether, 64.8 cents goes for military expenditure. The federal administrative machine eats up 18.2 cents, so that in the end 17 cents are left for our human needs. We have become slaves of our big military machines. Armaments have become the most lucrative business. Eighty billion dollars of our budget goes to the Pentagon. Our President vetoes relatively small sums going for health or education as inflationary, while shifting billions to armaments, lump sums like 2.5-12 billion going to the senseless ABM. Even the money we give to foreign aid seeps back to the Pentagon because we give it mostly to enable the underdeveloped nations to buy arms from us. We are driving the whole world into the insane and suicidal armaments race, to satisfy the ambition, craving for money and power of the military-industrial governmental complex.

What is the way out, if any? The first step is to clean out our government, chase out all the merchants of death and outdated hawks, and then make friends with our "enemies." Youth could have made an end to the Vietnam war long ago by tearing up all draft cards and looking upon anyone who accepts it as a traitor to mankind.

War and armaments are but one way to suicide. I will not discuss the other two, or three. I could not devote adequate space to them. Paul R. and Anna H. Ehrlich, in their admirable book, *Population, Resources and Environment* have dealt with it *in extenso* in a way which could not be surpassed.* All these ways to extinction have one common remedy: the development of a world-consciousness, the realization that we all belong to the same race. The way to happiness is not "dog eats dog" but "man is man's brother." Our little globe with its oceans, atmosphere and very limited resources is our common property and nobody has the right to pollute it, exhaust it or overpopulate it. The development of such a spirit is a

*W. H. Freeman and Co., San Francisco.

slow process. It demands a higher spiritual level than we have now and I only hope that we don't need a major catastrophe to reach it. We have long way to go, but even the longest journey begins with the first step.

Science has been so singularly successful because it has weights and measures which are the same everywhere. One gram or centimeter are one gram or centimeter everywhere. We have a "Bureau of Standards." What the new world needs is an "International Bureau of Moral Standards" to define what is good and what is bad. Good and bad are not abstruse ideas. Good is what is good for society, elevates life, does not hurt others; bad is the opposite. Killing, in itself, is not bad. Life is based on killing, the bigger fellow eating the smaller one. This is the law of the jungle, but a human society cannot be built on it. Killing was always regarded as the greatest of sins, but to satisfy military-industrial complexes and governmental ambitions we distinguish between two kinds of killing. The one is individual murder which is punished severely. The other is political or military murder, which goes unpunished and may even be looked upon as a kind of glory. There are two conditions for classifying murder as "political" or "military." One is that it has to be accompanied by some slogan or theory. The other is that it is not one man who has to be killed, but hundreds, a thousand, millions, or the whole or mankind. The higher the number, the more the killing is political or military.

In the new world of our shrunken little globe we will have to arrive at a more uniform definition. We will have to call murder killing a fellow man regardless of color, creed, passport or uniform, and regardless of the slogans, lies or theories which accompany it, and regardless of the number of the killed.

We would need some supernational organization to pass judgments. At present it is only the victor who judges the vanquished. Had the American Revolution failed, Jefferson, Washington and Franklin would have been hanged and gone down in history as criminals. I wonder how our Vietnam war and our political leaders would look to a supernational tribunal?

Mankind will find peace and safety when it reaches that level of moral and ethical development at which it looks at all instruments of murder with revulsion and the wearing of the uniform of the mass murderer's guild as a shame.

VI. AMERICA: WHAT TO DO?

Not the majority is silent; the administration is deaf.

George Wald.

The American Government is a very powerful institution. Without a good government nothing can be done, while with a good government all problems can be solved. So the first step towards a better future has to be the replacement of the present government by a better one, and the education of the public to enable it to elect such a government. What makes a bad government dangerous it not only its bad decisions but also its use of its educational power. The government is the most powerful educator of the people. A government of hawks invariable dehumanizes its people, making a change most urgent.

The Constitution provides for change; all we have to do is to find the voters who will vote for such change and help us replace outdated hawks with people capable of building a new world. It is for the young to convince the old that this is the only alternative to catastrophe.

If you want to convince somebody of your opinion, the first thing to do is to listen. Don't start by *telling your* opinion; rather, try to find out your adversary's. Show (and have) due respect for this opinion. By doing so you will gain your opponent's confidence instead of putting him on the defensive. At the same time you will find out the essential differences between the two points of view, the problems worth discussing for which you can produce good arguments. Try to place yourself into your opponent's position. Don't try to overwhelm him; start gently. Give your opponent the feeling of working together, of being your associate in search for the best solution. Don't hesitate to admit it if your opponent's argument is better than yours.

On the whole you will find that most people believe themselves to be right just as much as you do, and you must be understanding of their weaknesses. Many people are simply afraid of any change and want to keep the world as it is or make it as it was. They are afraid that any change, eventually, may lead to destruction, revolution or communism. They think that it is the road to the left which leads there. You must convince them that it is the road to the right, the rigid adherence to the past which leads to trouble. You must explain that we are living in a new age and things must be adjusted to it, and can be adjusted constitutionally, without bloodshed and destruction.

There is a tragic misunderstanding which stands in the way of an approach between youth and the older, power-holding generation that has to be eliminated. Misled by superficial signs, the older generation think that today's youth wants to destroy its world. What youth wants is to save it, clean it up, add to its content and adjust it to the new age before it destroys itself.

If you are talking to a woman who has children you can appeal to her motherly feeling, for no woman likes to see her children killed. If you succeed in convincing her, you may be able to make her an ally who will convince her friends.

I would start my work at home, convincing my father and mother, my brother and sister, then convince my friends and classmates. Then I would start going into the streets, refusing to become discouraged by failure or even ridicule. Bernard Shaw said that he who believes in something should be willing to go out on the street corner and turn somersaults if he can promote his cause with it. If you happen to have long hair and find that it estranges people, don't take it as an insult or a sign of narrowness. You would feel likewise in their stead, with their past. I would not even hesitate to sacrifice my long hair if I felt that it estranges people, and if bad came to worse, I would put on a jacket and shave. Keep a cool head, show understanding, respect and good will. Those who are on our side already, must be convinced that their opinion counts only if they vote.

All this may be slow and may not fit many youthful temperaments, but it is the only sensible way. Violence is no good. It is self-defeating. It is an admission of impotence, lack of clear thought and intelligence, and should be avoided. Violence leads only to disaster. It is letting off steam in the wrong place. It leads to destruction which is a loss for all. Problems are like equations, which cannot be solved by blows, and revolutions mostly end where they started after a great deal of suffering and devastation. Moreover, active violence creates backlash and frightens away many people whose votes are needed. Our government has members who try to excite violence. Don't fall into the trap which they bait freshly every day. Violence would discredit youth, lead to a takeover by the army and thus give them full power.

Marches are not very productive either. In spite of glib and soft words, our opponents are hawks, and hawks can act only as hawks, so while we march in an orderly way we can march till doomsday. If we get disorderly we become harmful. Marches may be necessary, now and then, to show strength and make us get together.

Youth needs no violence, it can be sure of its victory. It has powerful allies. The most powerful ally of youth is biology: all of us have to die, and old people die sooner than young ones. The difference is so great that it cannot be equalized by the old people's sending the young to be slaughtered in Vietnam.

There is also another biological rule, a rule without exception: everybody has a mother. Being a mother, bringing up a child, always entails a great deal of labor, time, and self-sacrifice. The links between mother and child are the strongest ties man knows, dominated by hormones stronger than iron chains, without which babies would die and mankind would cease to be. These ties may loosen later, but all the same mothers dislike to see their children killed. In addition to this, women have two qualities which make them especially valuable allies. One is that they have votes, which are just as effective as men's. The other is that they have more common sense than men. Men are easily carried away by imagination. By virtue of their ages-old vocation of bringing up children, women are more closely linked to reality. They have breasts, which remind them of their ancient duty of feeding their children, while their menstruation reminds them of another great reality, that it is they who have to bear children, carrying them through the first nine months inside their bodies.

Women are, on the whole, also more sensitive than men. Balzac compares a man making love to a woman to a gorilla playing on a Stradivarius. But women are especially sensitive now as they become aware of their inferior status and revolt against it. Men always tried to dominate them, as the stronger always tries to dominate the weaker, and it was only at the Church's Council of Trent in the XVI century that men voted a soul to women. Until then they did not even have souls. Modern life has made the superior muscle strength of men less important, and women are revolting. Youth should seek their alliance.

Youth must prepare itself for the difficult job of building a new world. Fifty-five per cent of Americans are under thirty and the leaders of the future are among them. There will be no need to put this country on a new foundation. The foundation is there: our Constitution. Only its letter has to be filled again with the spirit of Franklin and Jefferson and with an eye on the rules of the new age.

It is very difficult to convince anybody, after 40, of anything new. If you can show definitely that their opinion is wrong they do not change their mind but become angry and call you a "communist," which cuts short discussion. So don't get discouraged if you meet rigidity and opposition.

VII. THE DUALITY OF SOLUTIONS

Jerome Wiesner has pointed out that there is no purely *technical* solution for the armaments race. The same is true for most of our problems. Roots lie deeper. Attempts at technical solutions alone may even aggravate the problems; most governments know technical solutions only.

Our technical solution for the *armaments race* is to increase our arsenal, make more bombs and improve them. Any improvement in our arsenal will force our opponents to improve theirs, adding a new turn to the armaments spiral. The real solution is to make friends with our enemies (as advised by Lincoln). There would be no point in arming against a friend. The real and final solution is in a moral and cultural progress, in a world-consciousness. Real safety lies only in a spirit which makes war unthinkable.

One of our greatest problems at present is *drugs*. The technical solution pursued by our government is the passing of the famous "no knock" legislation, the employment of beefed-up narcotics squads, and the dealing out of severe punishment without discrimination between innocent and dangerous drugs.

The real remedy is a livable world, the restoration of the dignity of the individual and the value of endeavor.

The technical solution to *crime* is a stronger police force. Its real solution is the elimination of wars, slums, misery, poverty, hunger, ghettos, ignorance and unemployment. Our technical solution to *violence* is tear gas and the National Guard. Its real solution is understanding and good will and the elimination of passive violence, the rigidity of attitudes which will budge only to active violence, begetting it.

Science has become a threat to the existence of mankind. The technical solution is arresting its progress. The better solution is using science's wonderful results for the elevation of life, reduction of toil and suffering instead of converting them into instruments of death.

There is no doubt that a great deal of technical know-how will be needed to combat *pollution,* but here, too, technique will never be enough. Kaiserling told us that "whoever wishes to better external conditions must begin by improving the inner man." What we need is a feeling of human solidarity, to look upon the earth and its atmosphere as the common property of man, to be carefully guarded by everyone.

Our *racial problems,* too, have a dual solution. They require a great deal of technique, of course, such as in the problem of school integration, but the real solution to racial problems lies in human beings accepting one another as equals and eliminating the factors of racial difference as a measure of one's social or political rights.

The greatest and most acute problem of all is *war,* which can produce a worldwide catastrophe at any moment. The technical solution lies in a strong UN. The real and final solution is a spirit of human solidarity, the understanding on the part of all people of the world that we belong to the same race and depend on one another. So long as the present narrow nationalistic spirit prevails and we continue to attach more value to national rather than human sovereignty, there is little hope for us.

Technical solutions, in themselves, may not only be ineffective but may achieve the opposite of what they are designed to achieve — force creating counterforce.

Real solutions will demand money. But there *is* money. We are a tremendously rich country. The trouble is that we often use our wealth wrongly. With the money spent in Vietnam in one single day we could have twenty *S. S. Hope*'s* circling the world for a year. With the money spent there we could have been well on the way to solving all our problems, eliminating slums, hunger, poverty, ghettos, ignorance and crime. We could have raised the level of human life all over the world. We could have fulfilled the dream of democracy instead of discrediting it.

---

*Richard L. Tobin in *Saturday Review*, April 4, 1970.

S. S. Hope is a hospital ship bringing medical help to those who need it.

VIII. ON INSTITUTIONS

Like individuals, institutions too become senescent. In our age of rapid change this senescence is rapid. Being composed of individuals, the personal interests of whom are identical with those of the institution they serve, most public institutions try to achieve more power and end up serving chiefly their own self-interest instead of serving the purpose for which they were created. They tend to become masters, instead of remaining servants of the people. This has led to the hypocritical world that is now being rejected wholesale by our youth. Not only do institutions change, so does society. While the needs and philosophy which created the institutions change, the institutions themselves are left fighting for their archaic and vested interests. This trend of our institutions to dominate us is promoted by our having in us something of a "bicyclist" nature, i.e., the man who bends his back upwards and kicks downwards.

But how can I pass judgment on individuals or institutions? Who am I to judge others? But judge I must since, as a voter I have to say "Yes" or "No." Shall I listen to words, the main link between people? Shall I listen to promises, arguments, logic or philosophy? All these consist of words and words can be put together in many different ways. It was long ago that I made to myself the rule not to listen to any of these but follow an age-old advice, given by the bible: "you know the trees by their fruits." This is good common sense. It I see an apple on a tree I know it is an apple tree. If I see a pear, I know it is a pear tree. The good tree bears good fruit, so the Bible goes, and the bad tree bears bad fruit. The Bible goes even further and tells us what to do with the bad tree: cut it down. When judging people or institutions we should look at "fruits."

Our most important institution is the *Government.* We elect it to serve us and lead us to peace and elevate life. If it identifies itself with the country, treats honest dissent as treason, spies and keeps files on its citizens, polarizes society and, to use Jefferson's words, squanders the money of its people under the pretence of defending them, neglecting such priorities as health, education and social equality, it is no good and has to be felled. If it buys for our money poverty, crime, inflation, unemployment and unrest, if it divides our house against itself, then it has to be felled urgently. The trouble only is that once we elect a government we have to put up with it for four years.

It is difficult to judge without information, so the Government withholds information from its people; it withholds information even from its representatives, instead of making the people its partner by supplying information.

Frighted people are easier to lead; thus governments often find it handy to spread fear, hatred, hostility and distrust towards other nations instead of working for understanding and collaboration.

These are age-old governmental practices, and so are not especially frightening. However, if a government starts war, or extends it under the pretence of guarding the life of G.I.'s who would be perfectly safe at home, if it extends the misery of war to new countries, helping to scuttle their neutralist government, if it appeases public opinion by withdrawing from that invaded country only to have its army take the place of another army which is sent in with his bombers, adding double dealings to double talk, then the situation becomes serious and demands urgent action.

When its peace-making consists of creating a new army that fighting should never stop, when it goes into a conference of limitation of armament (Salt) by ordering a hasty expansion of its atomic arsenal by developing MIRV and ABM, when it spends more money on spying on his "enemies" than on exploring nature, when it scuttles UN, the hope of mankind, by taking the law in its own hands instead of submitting his differences to international arbitration, when it makes its country hated and detested instead of being loved, soils its flag by making it an ally of an antidemocratic, corrupt police government with its "tiger pits," when it degrades its country to the policemen of the world, when it dehumanizes its people with its "body counts," the number of killed human beings, to justify the slaughter of his own boys, makes a mockery of the sacred principles, as articulated by Jefferson, Madison or Hamilton, then something is basically wrong. When democracy means forcing on the neck of an underdeveloped people an autocratic government working with criminal methods, when deceit deceives and trickery tricks itself, then a major surgery is indicated.

This country always was the refuge of all the unjustly prosecuted. Now it fills other countries with its own refugees. A considerable fraction of its youth is being prosecuted and is filling its jails. This country used to attract the good scientists of the rest of the world while now its own scientists are fleeing it and so are many others searching a fulfillment of life. The government pats itself on the shoulder when the inflation and unemployment grow at a decreased rate while the country, with its enormous vitality, should be growing at top speed.

It is not possible to live on a small globe with hatred and fear in our hearts and megaton bombs in our hands. Never was there a greater need to unify mankind, underscore what is common in us and tone down what separates us, and never was mankind divided sharper in two camps. President Nixon is the master of dividing, polarizing, even our own house is divided, hawk from dove, young from old, white from black, and poor from rich.

The second most important institution is the *defense department,* which we maintain to defend us. If it takes our boys ten thousand miles away to defend us against poor underdeveloped country which never attacked us, uproots and devastates that people with its technical superiority, makes it into a nation of refugees, when it overgrows the government and becomes its master instead of being its servant, becomes the main adviser of government holding out the false carrot of final victory, when it converts all results of science into instruments of killing, defoliates virgin timber forest and crops by the millions of acres, conducting war against women and children who suffer first, when it makes society its slave taking away most of the fruits of its labor, and dominates even Congress and Senate, spreads fear and hatred towards other nations while spending of millions on its own "public relations," when it manages to fuse with industry and government until they become undistinguishable, as is the case at present in this country, then it's time for a thorough overhaul.

Analyzing the State Department, churches, or political parties, overall I would find the same self-service. Our basic principle should be that whatever we create should serve us, serve man, serve community, instead of becoming his master. Machines are fine as long as they are our servants, toil for us. If they become our masters, make life miserable filling the world with dirt and noise, then they are of no use. *Technocracy* is wonderful if it enables us to produce without sweat and provide us with more spare time for our human endeavors. If we can produce more efficiently under the sponsorship of big corporations, then let there be big corporations as long as they serve us, as long as they make life richer, not poorer. Technological progress is wonderful, but if it means fouling up our nest, making it ugly and uninhabitable, we should want no part of it.

This master-servant relationship also extends to our private lives. It is nice to have a big roomy home, well furnished with all the conveniences, but we must be careful that all these things belong to *us*, not that we belong to *them*. We must not fall for adverstising which tries to make us buy things we don't need with money we don't have.

The sum of all this is that whatever we create must be our servant and not our master, and must be judged by "fruits."

Many years ago I attended a national festival in Holland. The present queen, then a child, asked her mother, "Do all these people belong to me?" "No, *you* belong to *them*," was her mother's answer. Even a queen is only public servant No. 1.

IX. ON FOREIGN POLICY

Foreign policy can have but one aim: create peace, an equitable peace. If we believe that democracy is the way to it we must support democracy, without an edge against anyone.

One of the greatest errors or our foreign policy has been our support of dictators when it has been convenient to us. Cuba illustrates this point. I visited Cuba under the reign of Battista. He was our friend and ally, our shield against communism. I was shocked to see children, mostly in ill health, ill fed and dirty, all day long on the street. "Have you no schools?" I asked. "We have no schools," was the answer. "We pay taxes for schools but Battista takes the money away." "Have you no hospitals?" "We have no hospitals," was the answer. "We pay taxes for hospitals but Battista takes the money away." This was Battista, our great friend and ally. We liked him. He was easy to deal with. He could be bought cheaply.

Our government continues to cling to the simplistic illusion that the opposite of communism consists of dictators and military juntas, while in fact these are the doors for communism. Communists always vote extreme right. The opposite of communism is democracy. In our paranoic fear of communism we have been always ready to discredit democracy and ally ourselves with the devil himself, provided he was anti-communist. By

China we not only refused to sell grain to it for gold, we refused even to sell grain elevators to Canada which wanted to supply grain to China. Such actions reflect an inhumanity on our part that is impossible to logically justify. Would it not be worthwile, once, to attempt to be nice and understanding towards this huge nation to see how it reacts? Our military is already using it as a bogey-man to get the money it wants for its useless ABM. It might be worth a gamble to talk to China once as equals instead of building atomic bombs in defense against it.

We involved ourselves in the Vietnam war allegedly to prevent Chinese expansion. We went to war against Ho Chi Minh, who was our only shield against such an expansion; fighting him was analogous to fighting Tito to curb Stalin. Our "Vietnamization program" is based on Stalin's mathematics: when half of the people are for me, and half against, if I liquidate those who are against me, everybody will be for me. We declare zones that are not friendly to us "free fire zones," shooting at everything that moves within them. With the money spent there we could have given $3,000 to every Vietnamese, North or South, a sum sufficient to retire on. It would be difficult to decide whether the criminality or the insanity of this war is greater. It is high time to put a stop to it. President Nixon is afraid that we would be humiliated if we came away without a victory. We have already been profoundly humiliated; we humiliate ourselves every day again soiling our flag. Withdrawal would only ease this humiliation. He is also afraid of being the first President to lose a war; he does not notice that we are licked already, having been unable, with our enormous military might, to cope with a small underdeveloped nation. By being bogged down in Vietnam we have become a paper tiger; we have to stand by and watch quietly as the Soviet Union takes possession of the southern shores of the Mediterranean and the Suez Canal.

At the moment at which I am writing these lines in Woods Hole I should be in Stockholm where I was invited to participate at an international scientific conference. I had to resign my participation and lecture, feeling ashamed to go there, feeling to be a criminal who contributes to the maintenance of "Tiger Pits" in the prison on Con Son Island in Vietnam, who contributes to leaving a heroic, small and underdeveloped nation prey to our technically well developed but morally and mentally retarded army, against the will of that people which does not want us there.

As a scientist I have to sound a very serious warning. The most basic physical rule which governs the universe is that *entropy tends to reach a maximum, free energy a minimum.* In plain words this means that nature tends to randomize, to level off differences, to reduce everything to a lowest common denominator — reducing cities to rubble, wiping out cultures and civilizations. As expressed so succinctly by the great atomic physicist, Ernst Schrödinger, life is a struggle against this law.* We must be very careful not to allow our quarrels and bombs to work into the hands of entropy. We might become too successful.

*E. Schrödinger: What Is Life?, Cambridge University Press, Cambridge, England.

X. SCIENCE AND SOCIETY

Science is the dominating factor of our age. It is rapidly changing the face of human life. It has led us into the cosmotechnic age. It is the greatest contribution to and the leitmotif of the twentieth century. History will be made henceforth more in laboratories than in capitols. The fact that politicians can strangle science by withholding funds only proves their ignorance, not their superiority.

It is a tragic comedy that in this age of science there is not a single scientist in the Senate or Congress, and that our President treats science as one of the petty social luxuries. He does irreparable damage to it by withdrawing petty sums from science while dumping huge sums into the bottomless pit of war and armaments. We spend three times as much on spying on our enemies than on exploring nature.

The aim of science is to create new knowledge. New knowledge elevates life. Most often it leads to practical application; almost all the amenities and comforts of our daily life we owe to scientific discoveries. But even if knowledge finds no practical application, it elevates life. For most purposes it is quite irrelevant whether it is the earth which turns around the sun or the other way around. All the same, the acquisition of the knowledge that it is the earth which turns around the sun meant one of the greatest forward steps in human history.

What makes the importance of science sometimes hazy is the fact that science progresses by "basic discoveries," studies which throw light on the foundations of nature and which need varying lengths of time to bear practical fruits. It is also difficult, sometimes, to pinpoint the discovery which led to a practical development; it may have been the result of a long series of discoveries and improvements.

Within the American public at large there is a widespread revulsion towards science because it was science which produced the atomic bomb, nerve gas and defoliants, and most of the other devices that make our life miserable. This is rather an important point, and I would like to clear it up.

The main culprit, of course, is the atomic bomb, so let's put the record straight on this. It is we scientists who built it, but we built it not merely to have it, but to have it *before Hitler* did. Hitler had everything in hand he needed to build it. Had he had more faith in science than in astrology he could have built it, and it would have assured his world supremacy. We had to built it so as to get it before he did. When we finally had it, Germany had already collapsed and Japan was about to cave in. The scientific community, united under the Chairmanship of James Franck, demanded that the bomb not be used against Japan, that it should not be employed to destroy human life without compelling necessity. They proposed that we demonstrate its power to Japan on some uninhabited island. The military opposition, led by General Groves, prevailed, and the bomb was dropped, killing 200,000 people and opening an age of fear, jitters and balance of terror.

After the bomb was dropped the scientists proposed that its secret should be made public, the common property of man, under the condition that nobody would use it to destroy life. But the atomic bomb was kept secret, even from our allies, feeding the suspicions of Stalin that we wanted to use it against the Soviet Union. It was impossible to keep such a secret, and soon the Soviet Union knew all about it. It was the duty of its leaders to build bombs as fast as they could to balance the, in their view, American threat. This started up the arms race, which is still going on and will very likely put an end to "homo sapiens."*

---

*The tragic decision not to make atomic energy the common property of mankind and to keep the bomb a secret is an example of the fact that the combination of apparently insignificant circumstances can decide big issues, such as the fate of man. I want to preserve for history the following story:

Dramatis Personae: 1. *Winston Churchill*

2. *Sir Henry H. Dale,* leading British physiologist and pharmacologist who played an important, though anonymous, role in the British war effort. He was one of the great personalities who seem to be vanishing both from science and politics. He was a paternal friend of mine and I have the facts from him.

3. *Niels Bohr,* one of the greatest human beings who ever lived, a great man and a great scientist who played a decisive role in the development of the atomic theory and the atom bomb. Bohr spent an evening in my home at Princeton, where we both were members of the Institute for Advanced Study. He was a fascinating personality and he talked most of the evening, but when he left I had no idea of what he was talking about. He was a Dane and the Danish language makes a very poor mixture with English, and the English of Danes is often

Science produces tools, and any tool can be used both for construction and destruction. The more powerful a tool, the greater the destruction it can cause. Cosmotechnics has created enormously powerful tools which enable man either to build a new world with undreamt of wealth and dignity or to altogether wipe himself off the surface of the earth. Within our present political structure the scientist has little to say about the application of his discoveries. It is the military, a relatively small group of people, which converts the results of science into instruments of murder and destruction. So long as we spend more on this conversion, considerably more, than on science itself, so long as mankind prefers to support this conversion rather than support science and knowledge itself, we must remain a death-orientated society, one which will, eventually, get what it wants: death. The choice is ours.

---

difficult to understand without special training. I could not understand a word of his Danish-English. The same thing happened a few days later when I attended an informal lecture of his, sitting right behind him. I only knew from the program that he was talking about complementarity. Again I could not understand a word of what he said, but I was thoroughly fascinated by his wonderful personality.

The key figure behind the plan to make atomic energy and the secret of the bomb the common property of mankind was Churchill. He had to be convinced of the desirability of this course, and the scientists responsible for the development of the bomb in U.S.A. wanted to put their plan before him in writing. Bohr found the problem of such enormous significance to mankind that he insisted on taking the message personally to Churchill. Sir Henry Dale was instrumental in securing for him an interview of a full hour with Churchill, which was quite an achievement. When Bohr left, Churchill dryly remarked that he did not understand a single word of what Bohr said. This was the end of the idea which could have placed mankind on a more hopeful course.

The present situation is such a monumental idiocy that we have to go on with it because changing direction would be tantamount to admitting a mistake. Mistakes of this magnitude become tragedies when we fail to learn from them, but admitting a mistake demands great moral strength, and admitting a mistake is the last thing politicians want to do. They cannot stand to lose face, to risk damaging the image they have built of themselves, on which their elections or re-elections depend.

Science is a sitting duck for the politician who wants to show his deep concern for the taxpayer's money. Cutting the support of science is like cutting off roots in order to prune a tree. What makes science a sitting duck is the fact that most basic discoveries need time to bear fruit for everyday life; the damage done to science becomes evident only after it has become irreparable, *after* the next election. Without scientific research both industry and defense have to wither away. When I came to this country twenty-three years ago it was the leader in world science. Now the number of fields in which it has become second rate rapidly increases.

Science has not only doubled the span of human life, it is the watchdog of human welfare. It has probably saved mankind from one of the gravest calamities that was waiting for it with the exhaustion of the fossil sources of energy, oil, gas and coal. A significant part of the energy generated today is already produced by atomic power, and if we learn to control fusion, unlimited amounts will be available.

XI. GOALS AND PRIORITIES

If you want to get somewhere, the first thing you must know is where you want to be. What is our aim? What can be our priorities, the ultimate aim of human endeavor?

Our astronauts have brought home pictures of our little planet. There was no trace of human life visible on it. So if we were to exterminate ourselves, it would make no difference whatsoever. Our globe might even look prettier and greener without us, and there would be nobody beyond our world to shed a tear about our disappearance. Human life has no value in itself. It has as much value as we can give to it by being happy and decent, developing all that is good and beautiful in and around us. So men can have no higher goal than to elevate life, to create a world without hunger, fear and disease, a peaceful world dominated by decency, equity, good will, harmony and honesty, and led by wisdom and knowledge. We have gone a long way on this road and our national aim can be nothing other than to continue on it, enriched with the wonderful achievements of science and continually adapting our world to the requirements of the cosmotechnic age.

The question is: what to base this new world on? Some say love. Love is not enough as a foundation. Love is the result of a reaction between two individuals; it can easily turn into hatred. General love makes no sense; you cannot tell me to love my neighbor until you know my neighbor.

If there is a solid, sentimental foundation on which we can build such a world it is egotism — not the short-sighted egotism which sells the future for a fast buck, but a sophisticated one which knows its real self interest and understands that we all depend on one another; that by helping the retarded, the poor or underdeveloped, helping the whole, we can help ourselves. One of the most serious threats to the United States is the great, rapidly multiplying masses of poor and ignorant people. Soon we may find ourselves not envied, but hated and detested, an island in a hostile sea, cut off from the raw material on which our welfare and civilization depends.

Another foundation on which we can build a better new world, (without which no new better world can be built) is decency, good will, human understanding and solidarity, the knowledge that we are all members of the same race, depending on one another. We cannot all love each other, but we can respect each other. This, together, is, I think, what the youth calls "love", in political relations. I am deeply convinced that the best policy, in the long run, is fairness, good will, and honesty, while trickery and double talk defeat themselves.

Insofar as another foundation is needed, we have one: our Constitution and our Declaration of Independence. We only have to find the real leadership to fulfill the ideals contained in these documents. Lowest common denominators always have the greatest number, and so appealing to them is the easiest way to power. To power, but not to leadership. The qualities needed to get to power and to be a good leader are at present, mutually exclusive.

Man, the naked ape, does not even have fur to keep him warm. For millions of years he had a very difficult existence and had to struggle hard to feed his body and keep it warm. Only lately has science taught him how to produce without toil. This abundance is so new to us that we are now gobbling it up senselessly, wasting our irreplaceable resources, filling the world with our garbage. We exhaust ourselves in producing and consuming.

Material goods and high living standards are wonderful and worth striving for only insofar as we do not have them. Once we have them we find out that they cannot fulfill man's life. Our youth today hears the sound of "a different drummer," and distinguishes between price and value.

With increasing automation we soon will be able to produce much more than we can consume. What shall we do then? Senselessly pile up products, or lay off most workers? Sooner or later we will have to face the fact that we have to work less. But what are we to do with ourselves? How can we fill our lives meaningfully? This is the greatest problem of the future and we must prepare for it now, tearing ourselves from the fateful groove we are in at present, which is solving the problem by making war.

Human life is a strange and often paradoxical phenomenon. While we feel great and wise, a senselessly multiplying cell can cut short the span of our life with a terrible agony. Or else a little fibrin plug, formed in an artery of our brain or heart, can suddenly cancel our existence. These are the real humiliations for man, not admitting an error. Why spend all our means and ingenuity to show that our body can be torn apart by metal fragments or scorched by napalm? Why not put our means and ingenuity rather into eliminating the humiliation exercised by our natural enemies, so that all of us can live out our lives in decency and dignity.

We have arrived at a crossroads and have to choose which way to go. We can follow, under military and military-oriented leadership, our present road. We can continue to transform our world into a huge garrison, with all its customary drabness, moral turpitude and callousness; into a world in which endeavor and achievement have no merit; into a world of domination, exploitation and destruction; into a world of lies in which the military, in collusion with politicians, can tear young people from their studies and take them thousands of miles away from home to make them into killers or to have them killed. This road leads to the extinction of our race.

Or, we can rid ourselves of all these things, remnants of an earlier age, and create a world of moral, ethical and spiritual values, a world in which everyone has an equal chance to develop his or her talents to their highest degree. The inability to remember may be as disastrous as the inability to forget. Why not forget the outdated dark, barbaric, precosmic world and stretch out a friendly hand to everyone and follow the road towards a new world, a world the likes of which man has never seen? The best persuasion is a good example. The future does not start tomorrow. Every minute can be made into a new beginning.

XII. CONCLUSION

A Personal Note

The education of most of us begins before we are born. Mine started several generations before I was born, when a barefoot peasant boy took to the road in the poor northern part of Hungary to go south, where he hired himself to an apothecarian. Later he found himself Professor of Physiology and head of the department at the University of Vienna, then the center of "modern medicine." His son, and his son's son were all scientists. I am his great-grandson. According to my scale of values, established at an early age, the highest value worth striving for is knowledge, new knowledge, which elevates man. Early in my career I had to accept a great deal of deprivation and tribulation to serve this principle. Later Hitler taught me that knowledge, in itself, without moral values, is useless. He did not discourage me, for the problem was only how to get rid of him. Stalin did not discourage me either. He had a criminal mind but I knew that some day he would be gone. I came to the U.S.A., which was the country closest to my ideals. I hoped to be able to put my research experience into the service of the fight against cancer, which kills in this country alone one human every two minutes, mostly after prolonged agony, and which has deprived me of the two persons dearest to me.

At present, killing has precedence over healing, and owing to cuts in budget my research grant was recently cancelled, against the opinions of the scientific committee. Science is treated by the present government as a dispensable social luxury, and for the first time in my life I feel discouraged. I have had to ask myself whether I was wrong all the way? I have written down my thoughts to see them in black and white, on paper, so that I could have a sharp look at them and decide whether I followed a false trail.

My conclusion is that I was right. The speed with which my adopted and beloved country is going down hill proves to me that it is this country which has taken the wrong trail. I offer this short book to its youth in the hope that it will help put the nation's course right again. And I offer it to the entire country, which so generously welcomed me and provided for my research for many years.